SHOOTING STARS

BY SOPHIE WASHBURNE

Published in 2025 by The Rosen Publishing Group
in association with Scientific American Educational Publishing
2544 Clinton Street, Buffalo NY 14224

Library of Congress Cataloging-in-Publication Data
Names: Washburne, Sophie, author.
Title: Shooting stars / Sophie Washburne.
Description: Buffalo, NY : Gareth Stevens Publshing [2025] | Series: Scientific American investigates things that light up the sky | Includes bibliographical references and index.
Identifiers: LCCN 2024027960 | ISBN 9781725352261 (library binding) | ISBN 9781725352254 (paperback) | ISBN 9781725352278 (ebook)
Subjects: LCSH: Meteors–Juvenile literature. | Meteoroids–Juvenile literature.
Classification: LCC QB741.5 .W37 2025 | DDC 523.5/1–dc23/eng/20240702
LC record available at https://lccn.loc.gov/2024027960

Portions of this work were originally authored by Kristen Rajczak and published as *Shooting Stars*. All new material in this edition is authored by Sophie Washburne.

Designer: Claire Zimmermann; Andrea Davison-Bartolotta
Editor: Jennifer Lombardo

Photo credits: Cover, p. 1 AstroStar/Shutterstock.com; p. 5 Thomas Roell/Shutterstock.com; p. 7 Liliya Butenko/Shutterstock.com; p. 9 Triff/Shutterstock.com; p. 11 Craig Taylor Photography/Shutterstock.com; p. 13 XONIX/Shutterstock.com; p. 15 Adilson Honesto/Shutterstock.com; p. 17 The Courage to Travel/Shutterstock.com; p. 19 (bottom) Prystai/Shutterstock.com; p. 19 (top) AaronChenPS2/Shutterstock.com; p. 20 WeAre/Shutterstock.com; p. 21 VectorMine/Shutterstock.com.

Printed in the United States of America

CPSIA compliance information: Batch #CWSA25. For Further Information contact Rosen Publishing at 1-800-237-9932.

CONTENTS

Words in the glossary appear in **bold** type the first time they are used in the text.

FALSE STARS

Have you ever made a wish on a shooting star? It might surprise you to know that you were wishing on a meteor! The flashes of light they cause last for less than a second.

Each time a comet gets close to the sun, it melts and loses bits of itself. These bits are made of rock and other planetary materials. When they're floating in space, the bits are called meteoroids. "Meteor" is the word for both the chunk of rock and the streak of light it causes when it enters Earth's **atmosphere**.

Comets are balls of ice and dust that travel through space. From Earth, they look like bigger and brighter shooting stars.

FUN FACT

MOST METEORS COME FROM COMETS, BUT THEY CAN ALSO COME FROM ASTEROIDS. WHILE COMETS ARE MOSTLY MADE FROM ICE, ASTEROIDS ARE MOSTLY MADE FROM ROCK.

SHOOTING STAR STORIES

People have seen and studied meteors for hundreds of years. However, they didn't learn what meteors were until the mid-1800s. Before that time, people tried to come up with their own explanations for what they were seeing.

Some ancient **cultures** believed shooting stars were a sign from the gods. These could be either good or bad signs, but they were always seen as a signal that big changes were coming. Other cultures thought shooting stars were the souls of the dead.

In Norse **mythology**, shooting stars are made by the god Odin's eight-legged horse, Sleipnir. As Sleipnir moves across the sky, its hooves create sparks that can be seen from Earth.

FUN FACT

TODAY, MANY PEOPLE BELIEVE THAT A WISH MADE ON A SHOOTING STAR WILL COME TRUE.

SEEING THE LIGHT

Meteoroids that become meteors approach Earth with a lot of speed. **Friction** with gas **molecules** causes a meteoroid to slow down and heat up when it enters the atmosphere. The meteor loses some of its energy, or force, to the gas molecules. It releases some of this energy as light while the meteor glows and burns up. That's the light we see from a shooting star!

Most meteoroids that cause meteors are only about the size of a grain of sand. They can move as fast as 160,000 miles (257,440 km) per hour.

Because shooting stars move so fast, many people believe catching sight of one is a sign of good luck.

METEOR SHOWERS

Most meteors can be seen when Earth's **orbit** crosses the path of **debris** left by a comet's orbit. This path is called the meteoroid stream. Meteoroids from the stream move into Earth's atmosphere at high speeds.

Earth sometimes passes through a part of space with a lot of comet debris. More meteoroids in space mean more meteors in our skies! This light show is called a meteor shower. Meteor showers often happen annually, or at the same time every year, when Earth's orbit crosses the comet's path again.

During a meteor shower, it is possible to see up to 200 meteors per hour!

FUN FACT

METEORS THAT DON'T SEEM TO BE PART OF A SHOWER ARE CALLED SPORADIC METEORS. MOST NIGHTS, SPORADIC METEORS HAPPEN AS OFTEN AS EVERY 10 OR 15 MINUTES.

A YEARLY SHOW

Some annual meteor showers have been observed for centuries. They're often named for the **constellation** they appear to come from. However, meteor showers don't really start at constellations! These names just help observers know where in the sky to look for a shower.

The Perseid shower is one of the most famous. It commonly takes place between July 25 and August 18. Its meteors are debris from a comet called Swift-Tuttle.

FUN FACT

THE PERSEID SHOWER WAS FIRST RECORDED 2,000 YEARS AGO BY THE CHINESE.

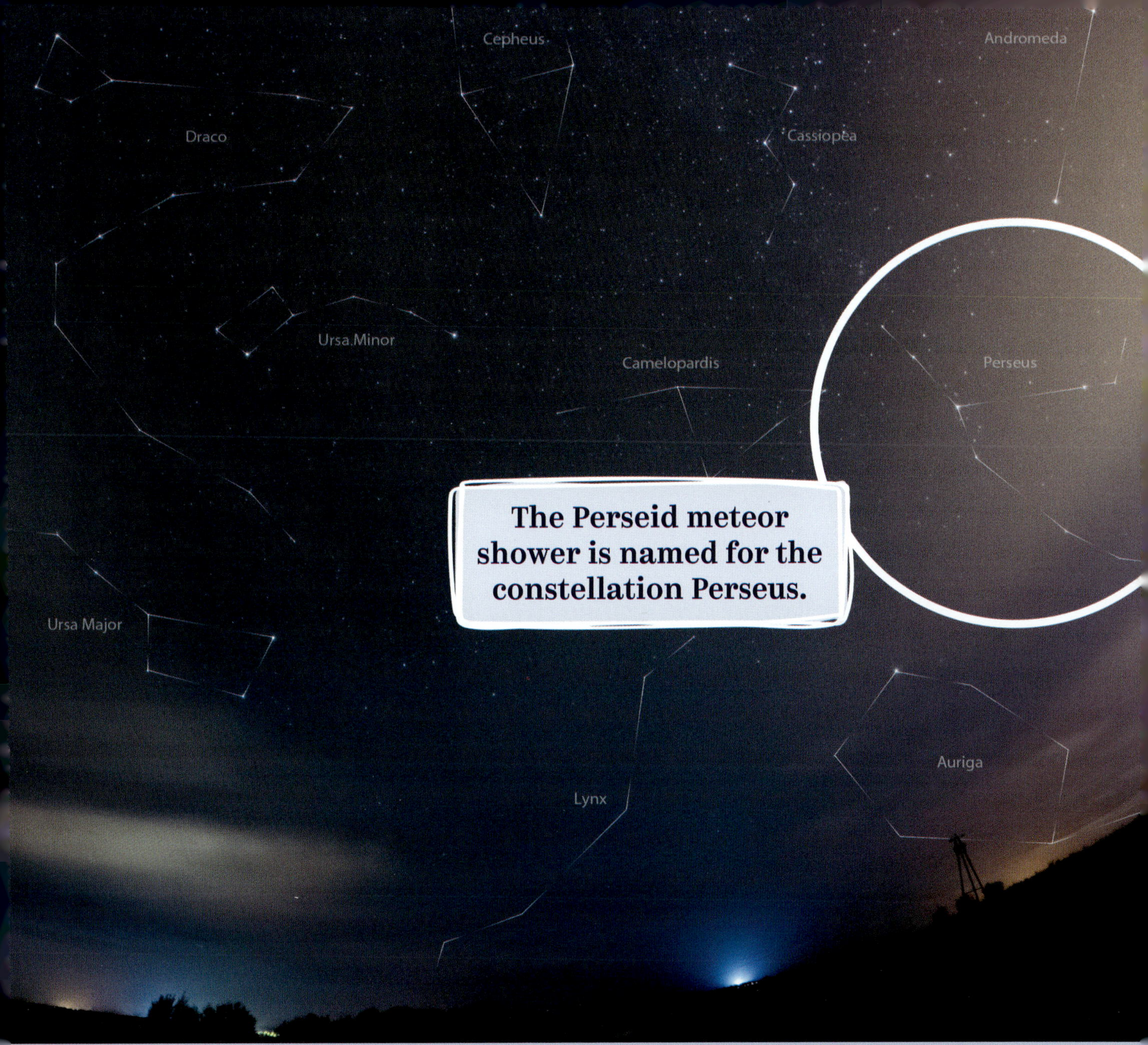

The Perseid meteor shower is named for the constellation Perseus.

MORE METEORS!

When meteors fall even more quickly and more often than in a meteor shower, it is called a meteor storm. To be called a storm, meteors must fall at a rate of at least 1,000 per hour.

Meteor storms are caused by Earth passing through a meteoroid stream with a lot of debris in it. These storms are very rare, or uncommon. There are about 30 meteor showers each year. However, a person may only see one or two meteor storms in their lifetime.

When people take pictures of the night sky over a long period of time, their camera can capture the movement of the stars. These star trails look like shooting stars. In reality, the stars aren't actually moving as fast as a meteor.

HITTING THE GROUND

Most meteors burn up in the atmosphere, so all that's left of them is a little bit of dust. However, if the meteor is big enough, part of it can fall to Earth. Meteors that reach Earth's surface are called meteorites.

The outside of a meteorite melts when it enters Earth's atmosphere, but it doesn't burn up completely. After it falls to Earth, a meteorite is cool enough to touch, though the outside will have been burned to a glossy black.

FUN FACT

THE STREAK OF LIGHT THAT SHOWS A METEOR'S PATH IN THE SKY IS CALLED THE METEOR TRAIL.

Most meteorites are very small. This picture shows an unusually large meteorite.

TYPES OF SHOOTING STARS

All shooting stars are really meteors. However, they don't all look exactly alike. Sometimes meteors can appear in different colors. This depends on what elements are in the meteor. When those elements burn, they give off different colors.

Most meteors appear high in the sky. Earthgrazers are meteors that appear low on the **horizon**—as if they're grazing, or barely touching, the ground. Next time you see a shooting star, notice where in the sky you see it.

The lights in a big city—or even a small town—can make an area too bright for stars to be seen. This is called light pollution. Stars are easier to see in places away from many buildings.

SEE FOR YOURSELF

It's possible to see one or two shooting stars just about every night, but only if you happen to be looking in the right direction at the right time. It's much easier to see them during a meteor shower.

When you hear about a meteor shower, plan to watch during the most active night of the shower. Find out where in the sky the shower will occur, so you can watch that area. Try to go someplace away from a city where there are very few lights.

This picture shows how a meteoroid turns into a meteor and then, sometimes, a meteorite.

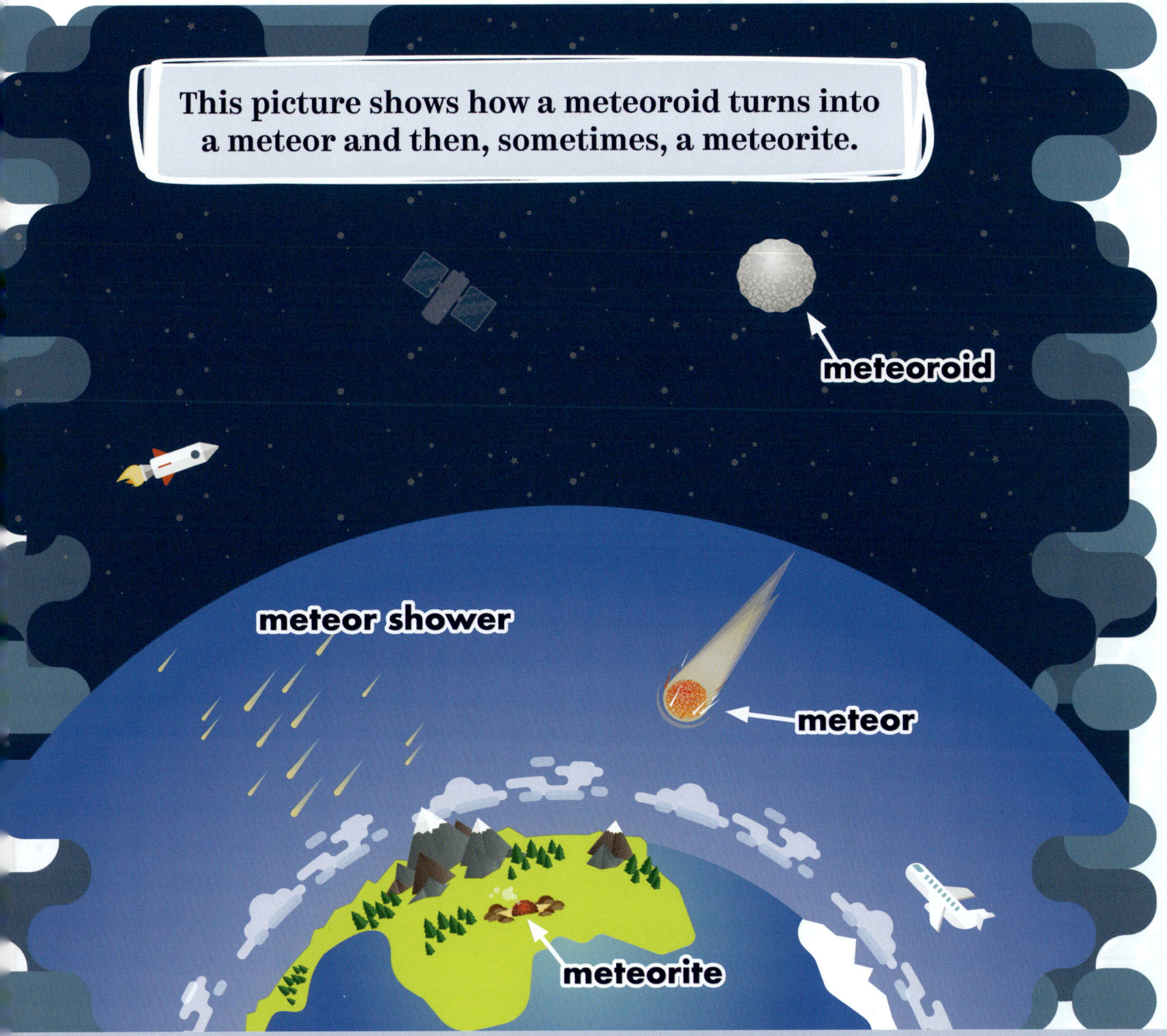

GLOSSARY

atmosphere: The thick layer of gases that surrounds the Earth.

constellation: A group of stars that forms a pattern.

culture: A group of people who share similar values and ways of life.

debris: The remains of something that has been broken.

friction: The force that slows motion between two objects touching each other.

horizon: The line where the earth or ocean seems to meet the sky.

molecule: A very small piece of matter.

mythology: A collection of stories that describe a culture's beliefs about themselves and the natural world.

orbit: To travel in a circle or oval around something, or the path used to make that trip.

FOR MORE INFORMATION

Books

Collins, Ailynn. *Tracking Meteors, Comets, and Asteroids with Velma*. North Mankato, MN: Capstone Press, 2023.

Harvey-Smith, Lisa. *Universal Guide to the Night Sky*. New York, NY: Thames & Hudson, Inc., 2024.

Websites

BrainPOP: Comets
www.brainpop.com/science/space/comets
Learn more about comets with Tim and Moby.

Kiddle: Meteor Facts for Kids
kids.kiddle.co/meteor
Read more about meteors and take a look at some cool pictures.

INDEX